The Adventures of
ARCHIBALD HIGGINS

HERE'S LOOKING AT EUCLID

(AND NOT LOOKING AT EUCLID)

IS THERE A MATHEMATICIAN IN THE HOUSE ?

The Adventures of
ARCHIBALD HIGGINS

Here's Looking at Euclid
Flight of Fancy
Computer Magic
Run, Robot, Run
The Black Hole
Big Bang
Everything Is Relative
The Silence Barrier

The Adventures of
ARCHIBALD HIGGINS

HERE'S LOOKING AT EUCLID
(AND NOT LOOKING AT EUCLID)

Jean-Pierre Petit

Translated by Ian Stewart

Edited by Wendy Campbell

William Kaufmann, Inc.
Los Altos, California 94022

Originally published as *Le Geometricon* © Belin 1980
published by Librairie Classique
Eugène Belin, Paris

Printed in the United States of America.

Library of Congress Cataloging in Publication Data

Petit, Jean-Pierre.
Here's looking at Euclid.

(The Adventures of Archibald Higgins)
1. Geometry—Popular works. I. Campbell, Wendy.
II. Title. III. Series: Petit, Jean-Pierre. Aventures
d'Anselme Lanturlu. English.
QA445.P4713 1985 516 85-5263
ISBN 0-86576-092-6

NOTICE

THIS IS NOT A TREATISE, OR A COURSE.

IT IS JUST A STORY OF ARCHIBALD HIGGINS
AND ONE OF HIS ADVENTURES
IN THE LAND OF GEOMETRY.

PREFERABLY TO BE READ WITH:

* PLENTY OF ASPIRIN

* AND LOTS OF STRING

* SOME SCISSORS
* STICKY TAPE
* A PROTRACTOR
* AND A NICE, PRETTY,
  ROUND BALLOON...

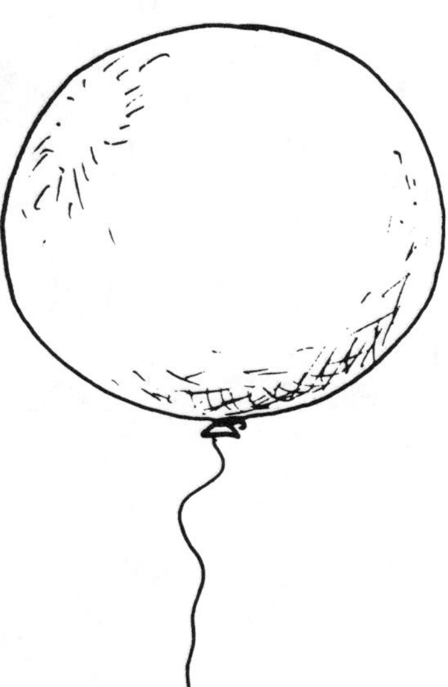

THE FIRM OF EUCLID & CO. WAS FOUNDED IN ALEXANDRIA IN THE THIRD CENTURY B.C. FOR TWO THOUSAND TWO HUNDRED YEARS THE BUSINESS PROSPERED. THE PRODUCTS WERE SUCCESSFUL, THE CUSTOMERS SATISFIED.

BUT, bit by bit, the customers' tastes changed. Some, who had previously never questioned the brand, after strange experiences, began to ask "Is Euclid **ALWAYS** the Truth, the Whole Truth, and Nothing But the Truth?"

HERE we recount the tale of one such person...

PROLOGUE: ONE DAY, ARCHIBALD HIGGINS WAS TRYING TO STRETCH A STRING BETWEEN TWO POSTS...

With three stretched lines, that is, three **GEODESICS**...

Archie made a **TRIANGLE.**

Placing his Protractor at each corner of the triangle, he measured the angles $\hat{A}$, $\hat{B}$, $\hat{C}$, and calculated their **SUM.**

Using an excellent theorem from the firm of Euclid & Co., this sum must be 180°. Good...

$$\hat{A} + \hat{B} + \hat{C} = 180^\circ$$ Ευκλιδ

ARCHIE'S HOME WORLD WAS COVERED IN THICK CLOUDS. YOU COULDN'T SEE YOUR HAND IN FRONT OF YOUR FACE.

BUT – AS YOU HAVE POSSIBLY NOTICED – THERE ARE DAYS WHEN NOTHING SEEMS TO GO RIGHT.

ARCHIE, HAVING PLENTY OF STRING, DECIDED TO CLEAR MATTERS UP ONCE AND FOR ALL.

QUITE UNDAUNTED, HE EXTENDED HIS STRING STILL FURTHER, ALWAYS **STRAIGHT AHEAD**, BURSTING WITH CURIOSITY.

SADLY...

I'VE GOT MY STAKE BACK AGAIN!

... ARCHIE'S **STRAIGHT LINE** CLOSED UP!

? ?
? ? ? ? ?
? ? ? ?
LET'S SEE WHAT THOSE EUCLID PEOPLE HAVE TO SAY ABOUT THIS. I'LL TRY DRAWING THREE GEODESICS OF EQUAL LENGTH, MAKING A TRIANGLE. THEN THE ANGLES SHOULD BE 60°, AND THEIR SUM 180°. JUST LIKE THIS LEAFLET SAYS.
B
60
60
60
A
C
Â+B̂+Ĉ = 180°
THEN WE'LL SEE.
RATIO OMNIA VINCIT.
A
B
A
HERE'S THE SECOND VERTEX B. NOW TO STRETCH TWO MORE LINES TO FIND THE THIRD.
ISN'T SCIENCE WONDERFUL?
B
OOPS! THE ANGLES ARE EQUAL ALL RIGHT - BUT THEY'RE BIGGER THAN 60°.
HIT A SNAG?
C
AND I HAVE A FEELING THAT MAKES THEIR SUM BIGGER THAN 180° !!
EUCLID
EUCLIDEAN GEOMETRY
EUCLID

AND YET, BY PLACING MY RULER ABSOLUTELY **FLAT** I CAN CHECK THAT MY LINES REALLY ARE **STRAIGHT**
EUCLID
HELLO? EUCLID & CO.? LISTEN, I'VE GOT PROBLEMS WITH YOUR PRODUCTS.
JUST A MOMENT ... I'LL PUT YOU THROUGH TO THE TECHNICAL DEPARTMENT.
PROBLEMS WITH OUR TRIANGLES? I'M SURPRISED. WHY DON'T YOU TRY OUR CIRCLES? OUR CUSTOMERS ARE VERY HAPPY WITH THOSE.
... RIGHT, GOTCHA. A **CIRCLE** IS A SET OF POINTS SITUATED AT A DISTANCE ℓ FROM A FIXED POINT.
AND YOU RECKON THE **PERIMETER** IS $2\pi\ell$ AND THE **AREA** $\pi\ell^2$. GOT IT!
ℓ
HAPPY TO OBLIGE.

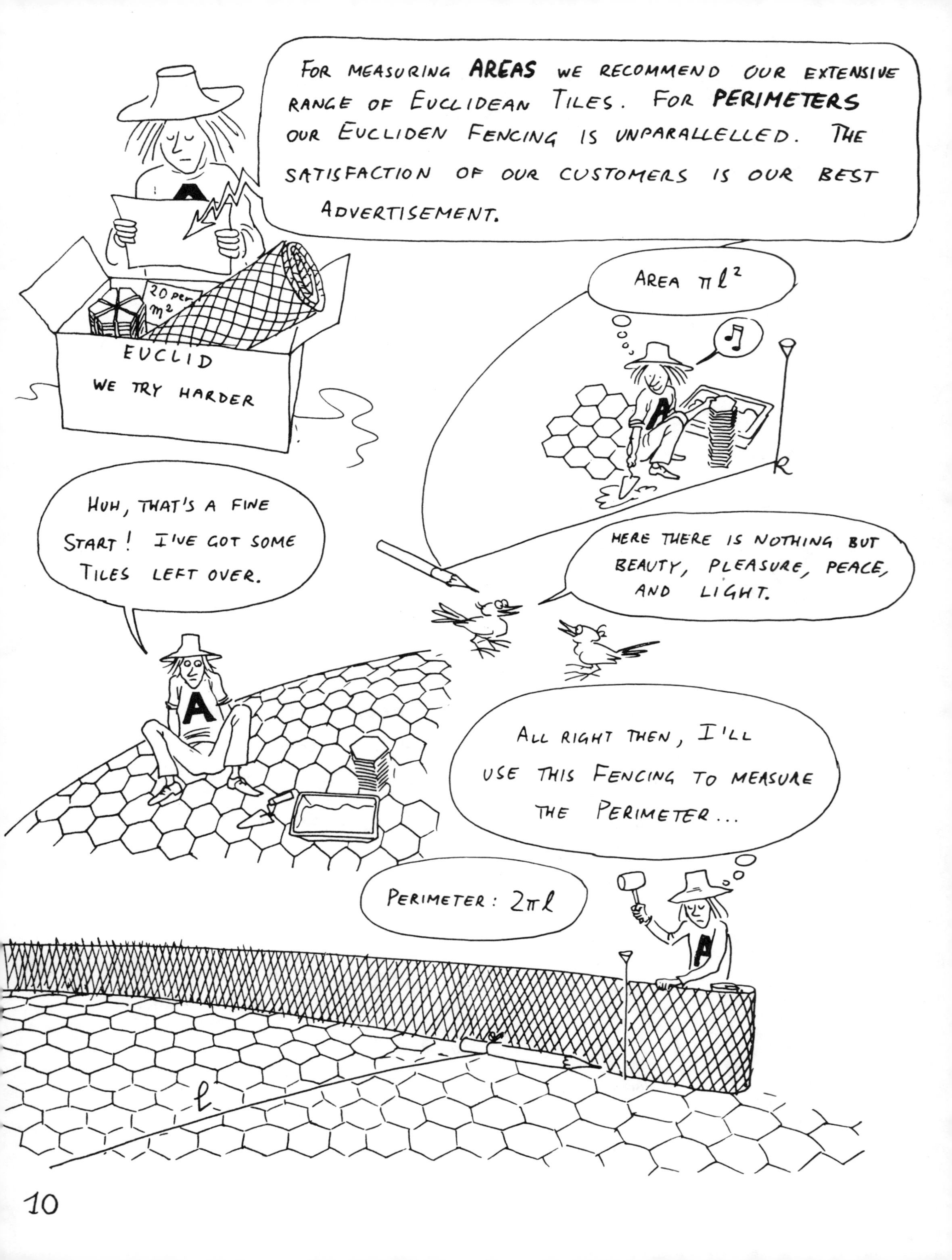

For measuring **AREAS** we recommend our extensive range of Euclidean Tiles. For **PERIMETERS** our Eucliden Fencing is unparallelled. The satisfaction of our customers is our best advertisement.
20 per m^2
EUCLID
WE TRY HARDER
Area πl^2
Huh, that's a fine start! I've got some tiles left over.
Here there is nothing but beauty, pleasure, peace, and light.
All right then, I'll use this Fencing to measure the Perimeter...
Perimeter : $2\pi l$
l

THERE'S SOME FENCING LEFT OVER TOO !!
HELLO, EUCLID & CO.? YES, IT'S ME AGAIN! I WANT TO MAKE A COMPLAINT ABOUT YOUR FENCES AND YOUR TILES! $\pi \ell^2$ AND $2\pi \ell$ AREN'T WORKING AT ALL! WHAT ARE YOU GOING TO DO ABOUT IT?
PLEASE DON'T SHOUT LIKE THAT, SIR. I'M ONLY THE SECRETARY. I'LL PUT YOU THROUGH TO OUR TECHNICAL DEPARTMENT.
NO! NO! THE TILES ARE PROPERLY JOINED! THERE'S NOTHING WRONG WITH MY RADIUS, AND THE FENCE IS PROPERLY ATTACHED TO THE CIRCLE!
SIR, PLEASE BELIEVE ME, IT'S THE FIRST TIME THIS HAS EVER HAPPENED. HAVE ANOTHER TRY, DON'T GET UPSET. YOU KNOW OUR THEOREMS ARE GUARANTEED.
ARCHIE CONTINUED HIS EXPLORATION, AT EACH STAGE INCREASING THE RADIUS ℓ OF HIS CIRCLE.
AND THE DISCREPANCIES GOT WORSE AND WORSE...

OH, LORD! NOW I'VE GOT OVER 36% TOO MUCH FENCING! AND 19% OF THE TILES LEFT OVER! AND THE CIRCLE I'VE DRAWN SEEMS TO BE A STRAIGHT LINE...
I MUST BE DREAMING, SURELY!
EUCL
WELL... IT LOOKS STRAIGHT ENOUGH!
ARCHIE INCREASED THE RADIUS FURTHER, AND NOW...
EUCLID
THE CIRCLE SEEMS TO BE BENDING THE OTHER WAY!
AND NOW, WHEN I INCREASE ℓ, THE PERIMETER GETS SMALLER! THIS IS CRAZY!
A

After yet more tiles:

WHAT HAPPENED?

To shed some light, let's blow away the fog...

Archie suddenly realized that he had been applying the rules of **PLANE GEOMETRY** while living on the surface of a **SPHERE**.

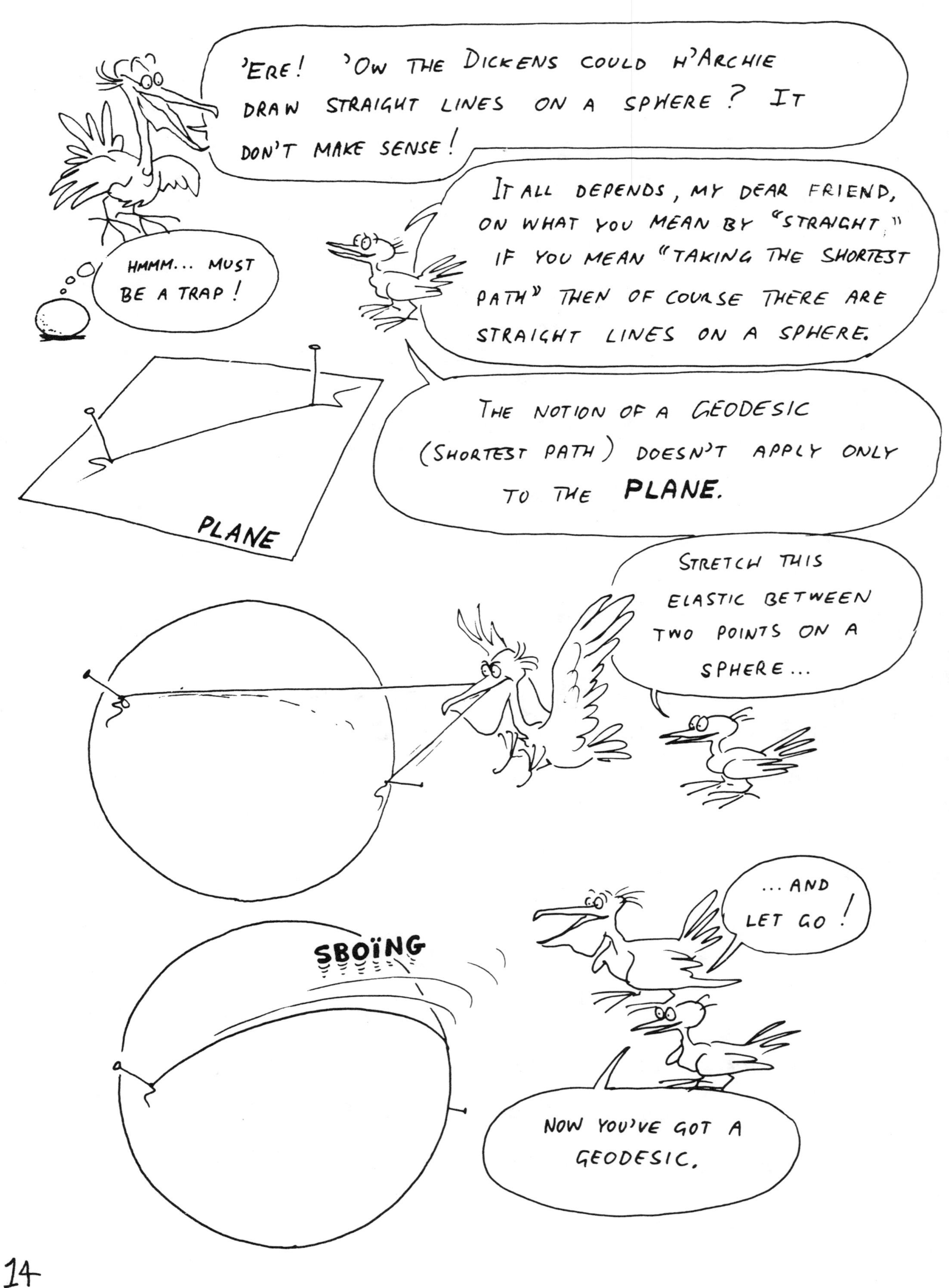
'ERE! 'OW THE DICKENS COULD H'ARCHIE DRAW STRAIGHT LINES ON A SPHERE? IT DON'T MAKE SENSE!
HMMM... MUST BE A TRAP!
IT ALL DEPENDS, MY DEAR FRIEND, ON WHAT YOU MEAN BY "STRAIGHT." IF YOU MEAN "TAKING THE SHORTEST PATH" THEN OF COURSE THERE ARE STRAIGHT LINES ON A SPHERE.
THE NOTION OF A GEODESIC (SHORTEST PATH) DOESN'T APPLY ONLY TO THE PLANE.
PLANE
STRETCH THIS ELASTIC BETWEEN TWO POINTS ON A SPHERE...
...AND LET GO!
SBOÏNG
NOW YOU'VE GOT A GEODESIC.

PULL THE UVVER ONE, IT'S GOT BELLS ON IT! THAT FING AIN'T STRAIGHT.
TAKE THIS RULER AND SEE FOR YOURSELF.
Y'CALL THIS A RULER ?
CERTAINLY. OF COURSE IT'S A RULER FOR SURFACES. ON A PLANE YOU HAVE TO USE THIS ONE. LOOK - IT BENDS NEITHER LEFT NOR RIGHT.
PLANE
IT'S STILL A DAM' FUNNY RULER.
RIGHT, NOW. WHEN THAT 'IGGINS BLOKE DREW THEM GEODESICS, THEY ALL CLOSED UP. DOES THAT MEAN THE GEODESICS ON A SPHERE IS JUST CIRCLES ?
EVERY LINE FOLLOWING THE SHORTEST PATH ON A SPHERE IS PART OF A CLOSED GEODESIC, WHICH IS A CIRCLE DRAWN ON THE SPHERE. BUT NOT JUST ANY OLD CIRCLE !

!???!
I FINK YER 'AVIN' ME ON, MATE! YER PLAYIN' WIV WORDS. YOU TRYIN' TER TELL ME THAT THERE ARE DIFFERENT KINDS OF CIRCLES ON A SPHERE? CRIPES!!
'ECK! I THOUGHT I UNDERSTOOD THIS STUFF – AN' NOW I FIND I DON'T UNDERSTAND A BLEEDIN' FING!
N
ℓ
A CIRCLE IS THE SET OF POINTS SITUATED A CONSTANT DISTANCE ℓ FROM A SINGLE FIXED POINT N, CALLED ITS POLE.
HUMPH!
N
HERE ARE A WHOLE LOT OF CIRCLES WITH THE SAME POLE N. WE CALL THEM PARALLELS.
S
ℓ
ℓ'
S
BUT THESE PARALLEL CIRCLES ARE ALSO THE SETS OF POINTS AT A CONSTANT DISTANCE ℓ' FROM THE "SOUTH POLE," ANTIPODAL TO N.
AMONG THEM, THERE IS ONE BIGGER THAN ALL THE REST – A KIND OF EQUATOR ON THE SPHERE.
LUMMY. I JUST SEEN WHY A CIRCLE ON A SPHERE 'AS TWO CENTRES N AND S!
THESE "EQUATORS" ARE KNOWN AS GREAT CIRCLES – AND IT IS THESE THAT FORM THE GEODESICS.
I NEVER SEEN A GEODESIC THIS CLOSE BEFORE. MOST IMPRESSIVE!

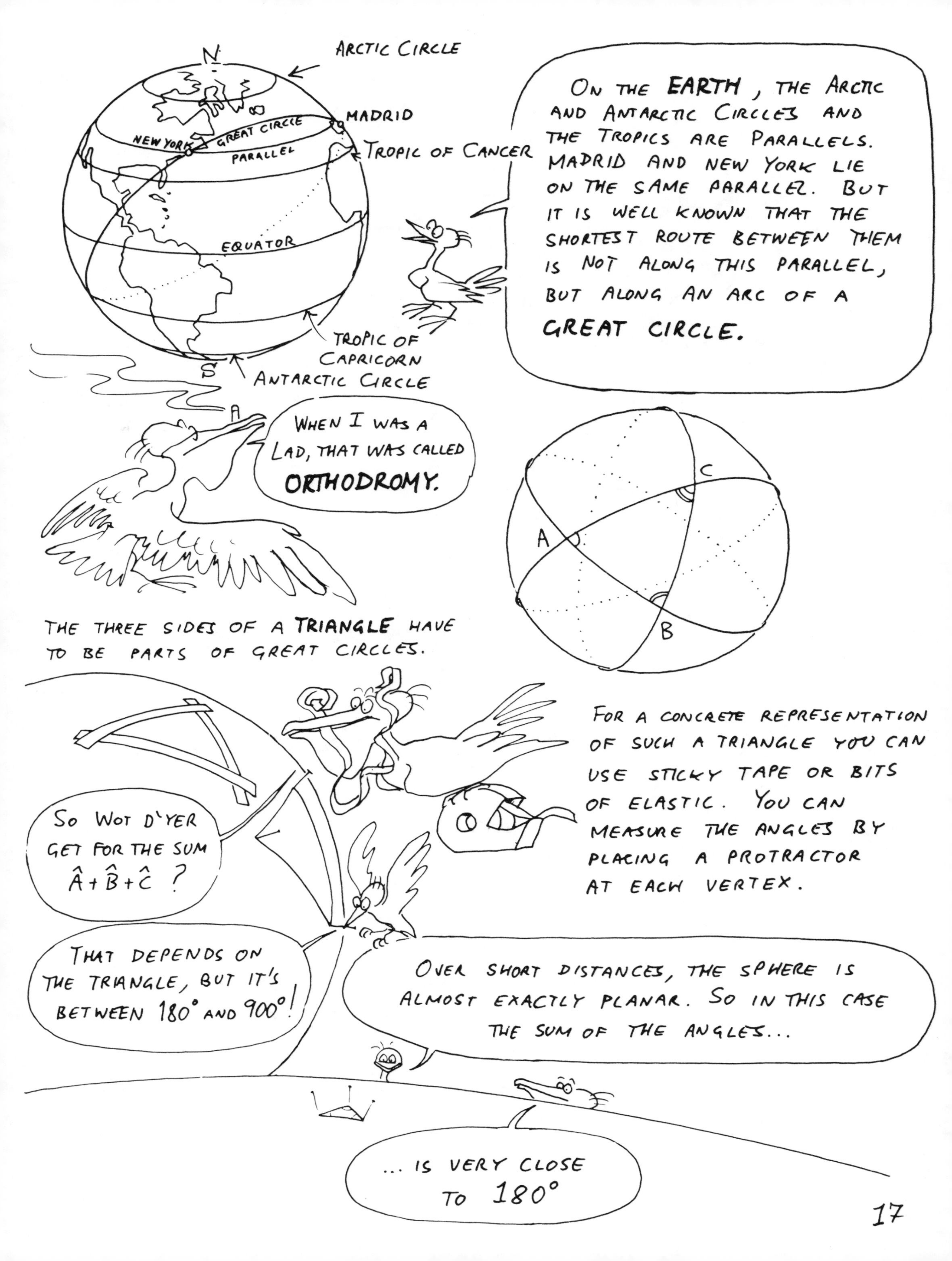
N
ARCTIC CIRCLE
MADRID
NEW YORK
GREAT CIRCLE
PARALLEL
TROPIC OF CANCER
EQUATOR
TROPIC OF CAPRICORN
S
ANTARCTIC CIRCLE
ON THE EARTH, THE ARCTIC AND ANTARCTIC CIRCLES AND THE TROPICS ARE PARALLELS. MADRID AND NEW YORK LIE ON THE SAME PARALLEL. BUT IT IS WELL KNOWN THAT THE SHORTEST ROUTE BETWEEN THEM IS NOT ALONG THIS PARALLEL, BUT ALONG AN ARC OF A GREAT CIRCLE.
A
WHEN I WAS A LAD, THAT WAS CALLED ORTHODROMY.
C
A
B
THE THREE SIDES OF A TRIANGLE HAVE TO BE PARTS OF GREAT CIRCLES.
FOR A CONCRETE REPRESENTATION OF SUCH A TRIANGLE YOU CAN USE STICKY TAPE OR BITS OF ELASTIC. YOU CAN MEASURE THE ANGLES BY PLACING A PROTRACTOR AT EACH VERTEX.
SO WOT D'YER GET FOR THE SUM $\hat{A}+\hat{B}+\hat{C}$?
THAT DEPENDS ON THE TRIANGLE, BUT IT'S BETWEEN 180° AND 900°!
OVER SHORT DISTANCES, THE SPHERE IS ALMOST EXACTLY PLANAR. SO IN THIS CASE THE SUM OF THE ANGLES...
... IS VERY CLOSE TO 180°

A
B
C
TRY MAKING A TRIANGLE LIKE THIS OUT OF STICKY TAPE OR ELASTIC.
90°
90°
90°
BLIMEY! H'IT'S A H'EQUILATERAL TRIANGLE AN' A THREE-SIDED RECTANGLE AT THE SAME TIME!
IT'S A VERY SPECIAL ONE - IT TAKES UP EXACTLY ONE EIGHTH OF THE SURFACE OF THE SPHERE.
AND THE SUM OF ITS ANGLES IS NOW $\hat{A}+\hat{B}+\hat{C}=270°$
!!?!
AND YOU - ER - "AIN'T SEEN NOTHIN' YET!"
IMAGINE A TRIANGLE, MADE OUT OF ELASTIC, WHOSE VERTICES MIGRATE OVER THE SPHERE. THE ANGLES GROW LARGER AND LARGER; AND SO DOES THE SUM.
N
180°!
THERE COMES A STAGE WHERE THE THREE VERTICES ALL LIE ON A SINGLE GREAT CIRCLE, THE EQUATOR OF THE SPHERE. THE ANGLES $\hat{A}$, $\hat{B}$, AND $\hat{C}$ ARE ALL STRAIGHT LINES, THAT IS, 180°. THEIR SUM IS NOW 540° !!

As the triangle continues its migration into the southern hemisphere, its vertices converge on the point S antipodal to N. Defining the vertex angles in the same way as at the start, they now exceed 180°! More precisely, they each become 360° - 60° = 300°.
S
360°-60°=300°
60°
SUM : 300 x 3 = 900°
A full circle comes to 360°.
Humph
300°
60°
So, on a sphere, the angle-sum of a triangle can lie between 180° and 900°!
In fact, a theorem proved by GAUSS says that the sum of the angles is given by:
Â + B̂ + Ĉ = 180 (1 + A / 3·1416 R²) degrees,
where R is the radius of the sphere and A is the AREA of the triangle.
SPHERICAL EOMETRY
When the area is small relative to the sphere, we recover the Euclidean result (Â+B̂+Ĉ = 180°)
A
B
C
If, on the other hand, the area of the triangle is almost that of the sphere, 4 x 3·1416 x R², we get 900°.

MEMORANDUM:

Two points of a sphere can be joined by two Geodesic arcs, making ONE Great Circle. But if these points N and S are ANTIPODAL, then INFINITELY MANY Great Circles pass through both! Two such lines on the sphere form a BIANGLE, with the same size of angle at each vertex. The angle sum can be ... ANYTHING !!

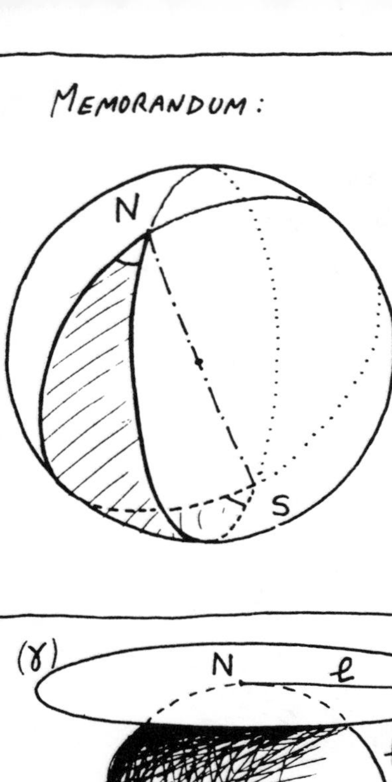

THEY'RE ALL MAD, Y'KNOW...

The Boss

(C) IS THE CIRCLE HE DREW, AND (γ) THE CIRCLE HE **THOUGHT** HE WAS DRAWING. FOR THE AREA, HE USED A FORMULA FROM PLANE GEOMETRY: $\pi \ell^2$ ($\pi = 3{\cdot}1416\ldots$). THE TRUE AREA IS HALF THE AREA OF THE SPHERE, $2\pi R^2$. NOW ℓ IS A QUARTER OF THE SPHERE'S CIRCUMFERENCE, $\frac{1}{2}\pi R$. SO THE RATIO OF THE TWO AREAS IS $\frac{\pi^2}{8} = 1{\cdot}233$. THE RATIO OF THE PERIMETERS IS $\frac{2\pi\ell}{2\pi R} = \frac{\pi}{2} = 1{\cdot}57$. IF YOU STILL DON'T BELIEVE ME, TRY WRAPPING THE DISC ON THE SPHERE!

DISC? DISC? **WHAT** DISC?

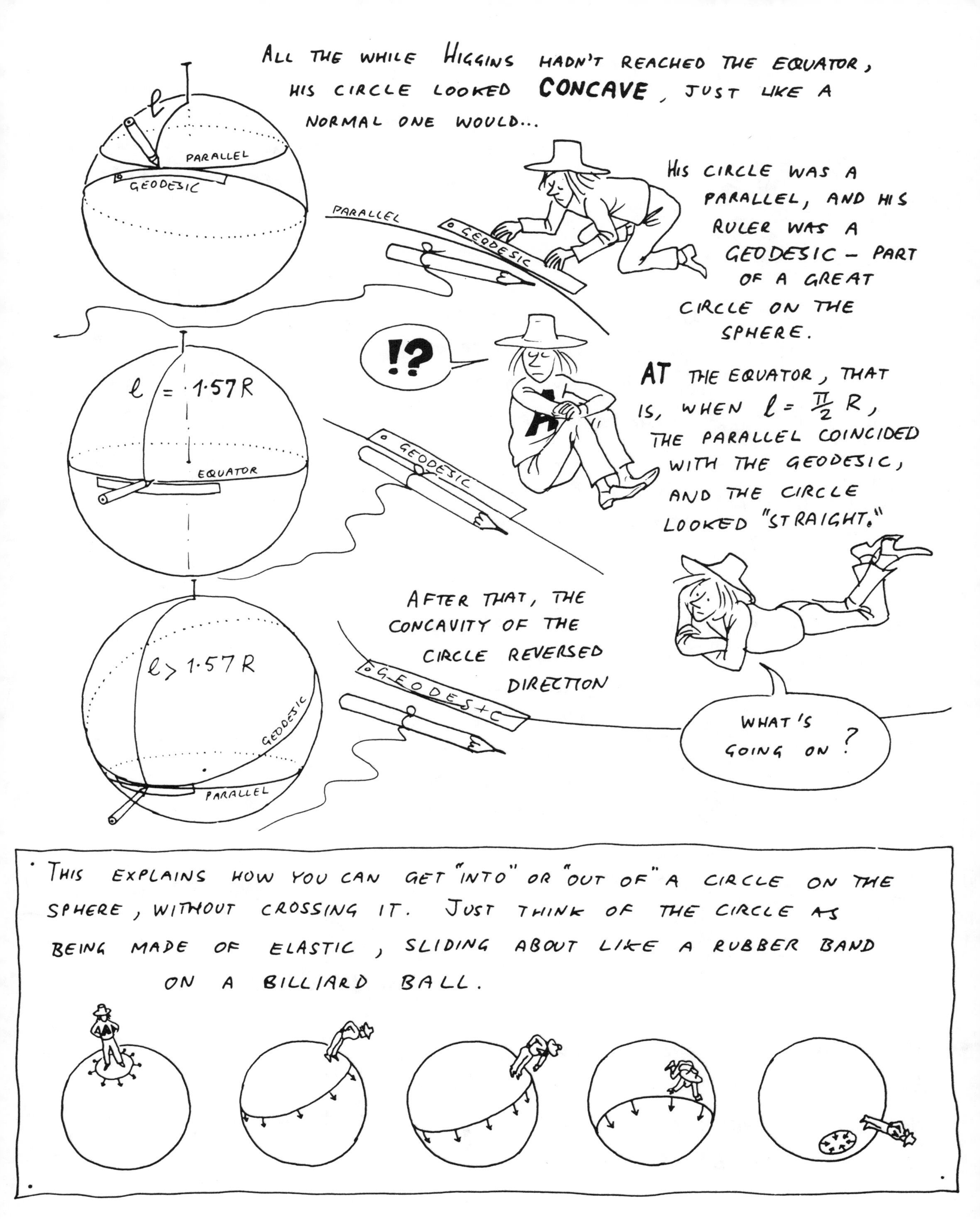
All the while Higgins hadn't reached the equator, his circle looked **CONCAVE**, just like a normal one would...
PARALLEL
GEODESIC
PARALLEL
GEODESIC
His circle was a parallel, and his ruler was a geodesic – part of a great circle on the sphere.
!?
$\ell = 1.57R$
EQUATOR
GEODESIC
AT the equator, that is, when $\ell = \frac{\pi}{2} R$, the parallel coincided with the geodesic, and the circle looked "straight."
$\ell > 1.57R$
GEODESIC
PARALLEL
After that, the concavity of the circle reversed direction
GEODESIC
What's going on?
This explains how you can get "into" or "out of" a circle on the sphere, without crossing it. Just think of the circle as being made of elastic, sliding about like a rubber band on a billiard ball.

HAVING REACHED A NEW WORLD, ARCHIE ONCE MORE UNREELED A GEODESIC — BUT THIS TIME...

?!
As always, defining a circle to be a fixed distance from a chosen point, Archibald Higgins found that a circle drawn on the new surface had a perimeter LARGER than 2πl, and an area GREATER than πl².
GET RID OF THE FOG :
The surface now has the same shape as a mountain pass, or the SADDLE of a horse. Many objects in daily use will serve equally well – a hunting horn, this sort of stool, or...
B
A
C
PLANE
B
A
C
I'd fall off THAT fing, mate !
No you wouldn't.
A
B
C
For the final word on all this, turn the page...
EUCLIDEAN GEOMETRY

CURVATURE:

A **curved** surface is one on which the theorems of EUCLID & Co. don't work. The curvature can be positive or negative.

On a surface of **positive curvature**, the sum of the angles of a triangle is greater than 180°. If you draw a circle of radius ℓ, its area is less than $\pi\ell^2$ and its perimeter is less than $2\pi\ell$.

On a surface of **negative curvature** the sum of the angles of a triangle is less than 180°. If you draw a circle of radius ℓ, its area is greater than $\pi\ell^2$ and its perimeter is greater than $2\pi\ell$.

A while back, Archie noticed that when you try to **wrap** a piece of the plane on a surface of positive curvature, pleats form in it. It is also impossible to wrap a piece of the plane on a surface of negative curvature: it splits.

This wrapping property is the simplest test for positive or negative curvature.

As you saw on the previous page, some surfaces can have regions of positive curvature **AND** regions of negative curvature.

I wonder if **CYLINDERS** or **CONES** are curved?

THE WRAPPING TEST...
HMMPH. YOU CAN WRAP A PLANE ON A CYLINDER OR CONE!
DON'T PANIC. I'LL STICK THREE BITS OF ELASTIC – GEODESICS – ON TO THE CYLINDER, USING STICKY TAPE.
B
A
C
... NOW, MARK THE GEODESICS ON THE SURFACE...
...ROLL THE CYLINDER OUT FLAT...
!
B
A
C
ACCORDING TO OUR DEFINITION, CYLINDERS AND CONES OBEY EUCLIDEAN GEOMETRY, AND SO ARE FLAT SURFACES !!!

THE NOTION OF SPACE :

EARLIER, THE CLOUDS PREVENTED ARCHIE FROM SEEING BEYOND THE END OF HIS NOSE – OR THEREABOUTS. WERE IT NOT FOR THAT, HE WOULD HAVE SEEN THE CURVATURE OF THE **SPHERICAL SPACE** HE LIVED **ON**.

THERE IS ANOTHER WAY TO PREVENT HIGGINS **SEEING** THE CURVATURE OF THE SURFACE : MAKE HIM LIVE **IN** IT – TO BE A **PART** OF IT.

NOTE THAT THIS NEW POINT OF VIEW HAS NO EFFECT ON:

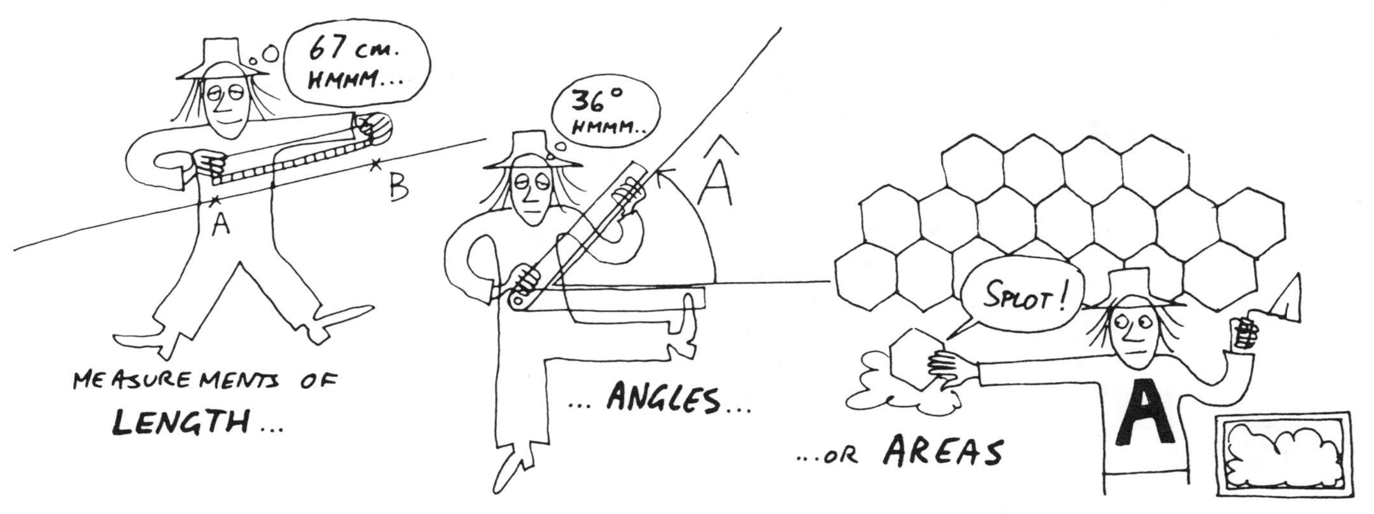

BUT, DESPITE BEING CONFINED WITHIN THE SURFACE ITSELF, ARCHIE COULD STILL CONSIDER ITS CURVATURE AND DECIDE WHETHER IT WAS POSITIVE OR NEGATIVE, AND EVEN MEASURE IT, WITHOUT BEING ABLE TO **SEE** IT. IF THE ANGLE-SUM OF A TRIANGLE WAS 180°, THE SURFACE WOULD BE A **PLANE**. IF THE SUM EXCEEDED 180°, THE CURVATURE WOULD BE POSITIVE, AND ARCHIE COULD CALCULATE THE **LOCAL RADIUS OF CURVATURE** R BY USING THE FORMULA $\hat{A}+\hat{B}+\hat{C} = 180\left(1 + \frac{A}{3.1416\,R^2}\right)$ DEGREES, WHERE A IS THE AREA OF THE TRIANGLE.

IF THE SUM WERE LESS THAN 180°, WE COULD DEFINE A RADIUS OF CURVATURE R GIVEN BY $\hat{A}+\hat{B}+\hat{C} = 180\left(1 - \frac{A}{3.1416\,R^2}\right)$, BUT IT WOULD NO LONGER HAVE THE **USUAL PHYSICAL MEANING.**

NOTE THAT WE CAN INCLUDE THE **PLANE** AS A SURFACE WHOSE RADIUS OF CURVATURE R IS **INFINITE**. BY SO DOING, WE WOULD RECOVER THE USUAL EUCLIDEAN THEOREMS.

THE CONCEPT OF DIMENSION

THE NUMBER OF DIMENSIONS IS JUST THE NUMBER OF QUANTITIES — OR **COORDINATES** — THAT MUST BE GIVEN, IN A CHOSEN SPACE, TO DEFINE THE POSITION OF A POINT.

SURFACES ARE SPACES THAT HAVE TWO DIMENSIONS. THE QUANTITIES USED FOR THE MEASUREMENTS CAN BE LENGTHS, NUMBERS, ANGLES...

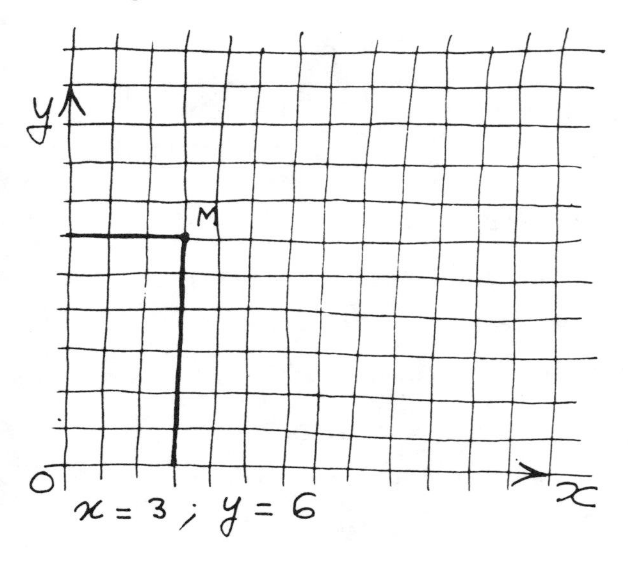

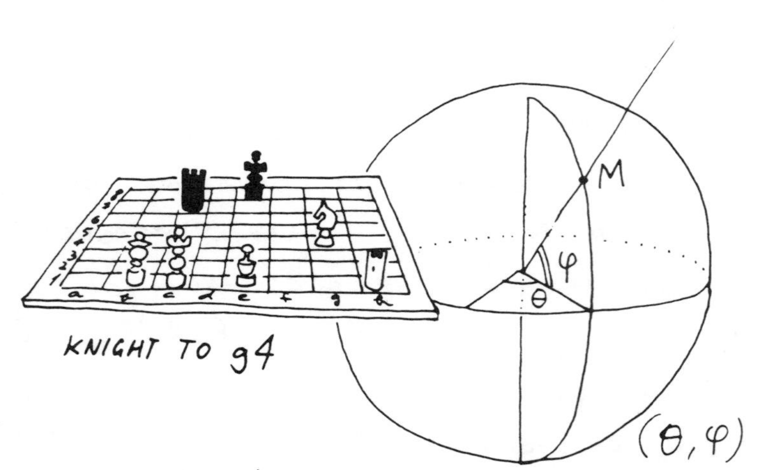

IT IS CUSTOMARY TO SAY THAT OUR SPACE, IF ONE IGNORES TIME, HAS 3 DIMENSIONS.

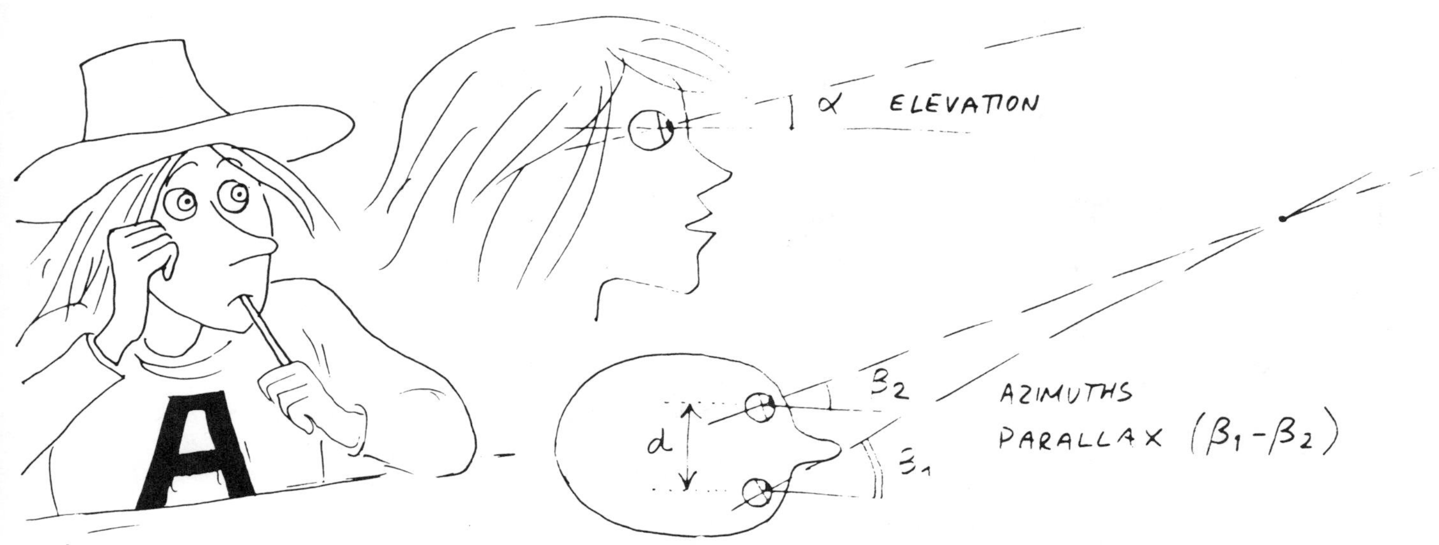

ARCHIE CAN FIND THE POSITIONS OF THINGS BY USING HIS SKULL...

THE POSITION OF A POINT CAN BE DETERMINED BY THREE **ANGLES**: THE ELEVATION α, AND THE AZIMUTHAL DEVIATIONS β_1 AND β_2 OF HIS TWO EYES.

THE ANGULAR DIFFERENCE $\beta_1 - \beta_2$ IS CALLED THE **PARALLAX.**

ARCHIE'S BRAIN CAN DECODE THIS PARALLAX, AND INTERPRET IT AS DISTANCE.

IMMERSION:

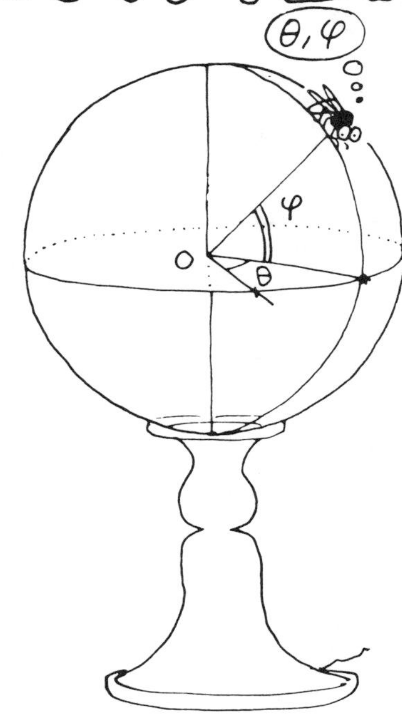

BUT THE FLY THINKS OF HIMSELF AS MOVING ON THE SPHERICAL LAMPSHADE, WHERE ITS POSITION, IN THIS 2-DIMENSIONAL SPACE, CAN BE DESCRIBED BY ONLY TWO ANGLES θ AND φ (LONGITUDE & LATITUDE).

WE SAY THAT THIS 2-DIMENSIONAL SPACE IS **IMMERSED** (OR **EMBEDDED**) IN OUR USUAL 3-DIMENSIONAL SPACE.

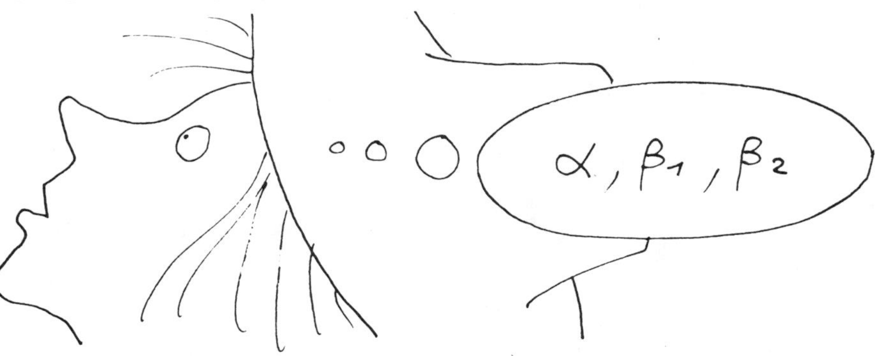

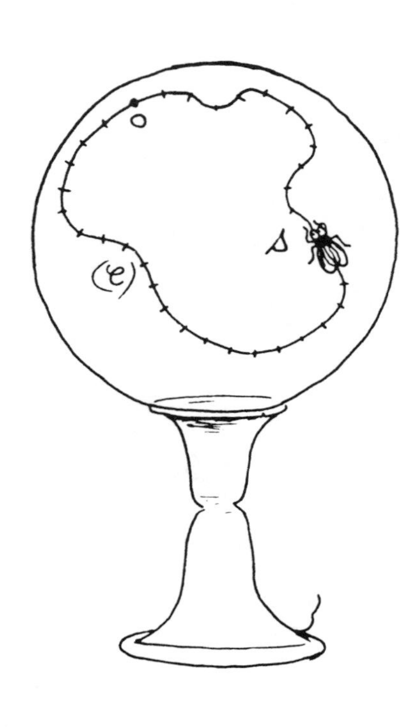

SUPPOSE THE FLY FOLLOWS A CURVE (𝒞) ON THE SPHERE. NOW WE CAN REPRESENT ITS POSITION USING ONLY **ONE** COORDINATE — THE DISTANCE FROM THE STARTING POINT (TAKING BACKWARDS DISTANCES AS NEGATIVE). A CURVE IS A PICTURE OF A 1-DIMENSIONAL SPACE.

THIS 1-DIMENSIONAL SPACE IS IMMERSED IN A 2-DIMENSIONAL SPACE (THE SPHERE) WHICH IS ITSELF IMMERSED IN A 3-DIMENSIONAL SPACE. SO OUR OWN SPACE **COULD** ITSELF BE IMMERSED IN ONE OF HIGHER DIMENSION, OF WHICH WE ARE NOT CONSCIOUS.

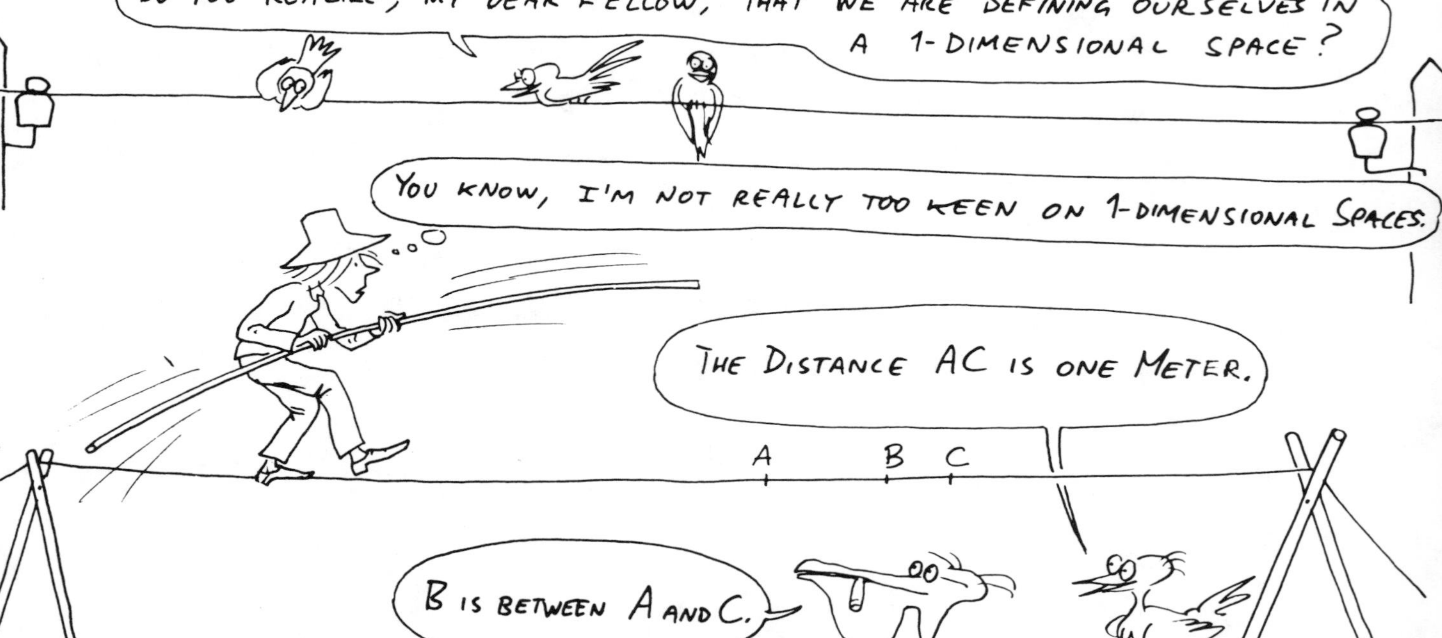

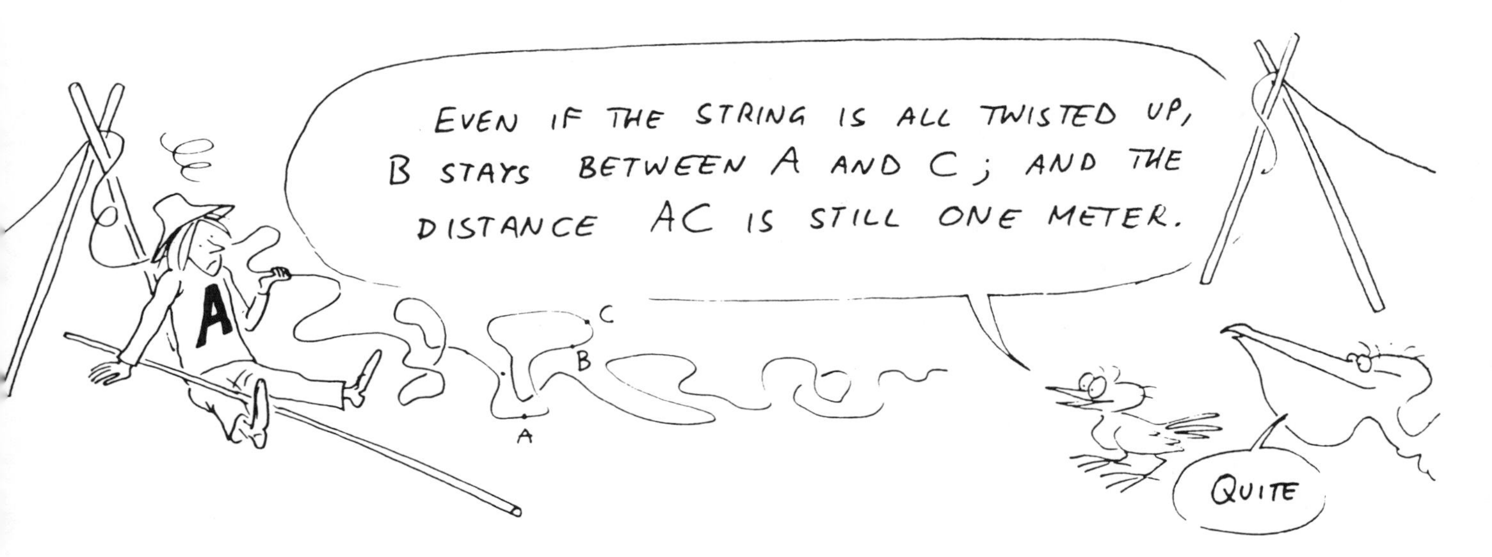

THIS SUGGESTS THAT SOME PROPERTIES CAN BE INDEPENDENT OF THE MANNER IN WHICH THE SPACE IS IMMERSED.

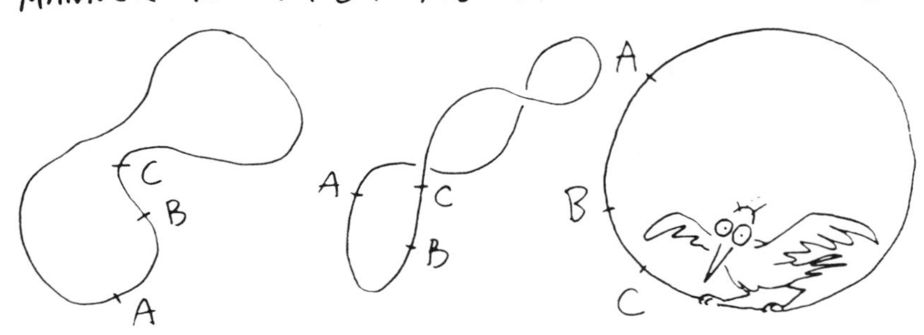

HERE ARE DIFFERENT WAYS TO IMMERSE A **CLOSED CURVE** IN ORDINARY SPACE. THE FACT THAT IT IS CLOSED DOES NOT DEPEND ON HOW IT IS IMMERSED.

BUT WE DO HAVE TO BE CAREFUL NOT TO STRETCH OR COMPRESS THE STRING, SO AS NOT TO CHANGE THE **DISTANCE** BETWEEN POINTS. NOW LET'S TRY IMMERSING **SURFACES** IN ORDINARY SPACE.

IF WE IMMERSE A **PLANE** IN ORDINARY 3-DIMENSIONAL SPACE, WE CAN BEND IT WITHOUT ALTERING ITS **INTRINSIC GEOMETRY**.

WE'VE SEEN THAT BENDING A PLANE INTO A CYLINDER DOESN'T ALTER GEODESICS OR ANGLES.

FROM THIS POINT OF VIEW A WAVY SHEET ALWAYS HAS A **PLANE EUCLIDEAN** GEOMETRY.

AN INHABITANT OF SUCH A TWO-DIMENSIONAL SPACE WOULD HAVE NO IDEA OF THE TWISTS AND TURNS AND UPS AND DOWNS OF THE SURFACE, WHICH ARE MERELY VARIABLE FEATURES OF THE WAY THE SURFACE IS IMMERSED IN 3-DIMENSIONAL SPACE.

IT'S CONCEIVABLE THAT OUR USUAL 3-DIMENSIONAL SPACE COULD BE IMMERSED IN ONE OF HIGHER DIMENSION, WITHOUT US REALIZING IT.

SUCH AN IMMERSION WOULD NOT CHANGE GEODESICS, NOR OUR PERCEPTION OF THE WORLD, BASED ON RAYS OF LIGHT WHICH MOVE ALONG GEODESICS.

A

HEY PRESTO!

B

WHICH MEANS WE CAN VISUALIZE THE POSSIBILITY OF A PATH BETWEEN TWO POINTS, SHORTER THAN THAT TAKEN BY LIGHT.

YER DON' SAY!

I KNOW WHAT YOU'RE UP TO! YOU'RE TRYING TO GET ME INVOLVED IN SCIENCE FICTION!

WHAT **ARE** YOU DOING?

EXPLORING THE END OF MY SHELL.

TAKE A PIECE OF THE PLANE AND FOLD IT:

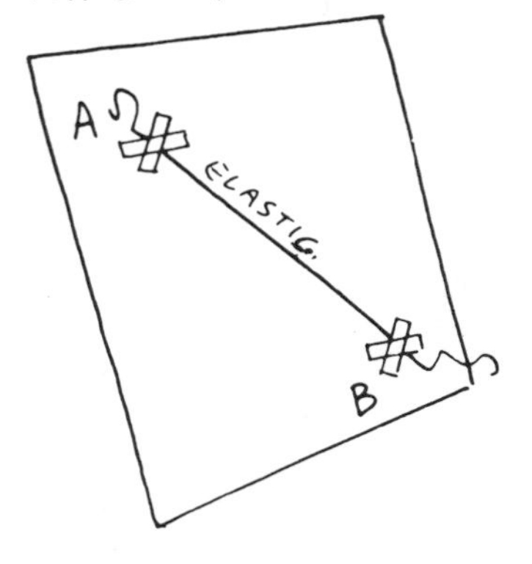

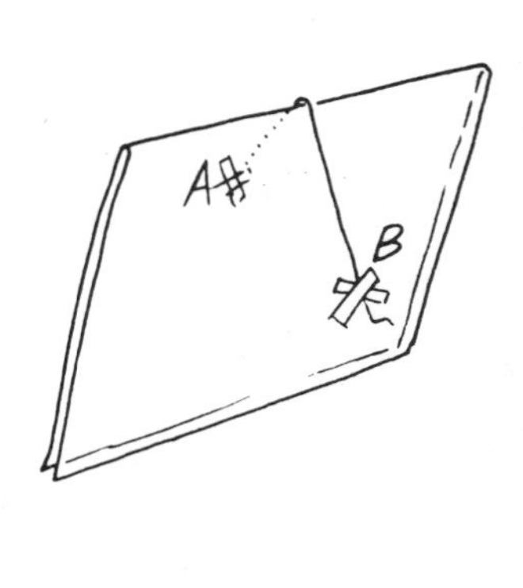

THE FOLD DOESN'T ALTER THE PATH OF THE GEODESIC AT ALL!

USING A RULER, DRAW LOTS OF STRAIGHT LINES (GEODESICS) ON A SHEET OF PAPER. THEN FOLD THE PAPER SEVERAL TIMES. THERE, BEFORE YOUR VERY EYES, ARE THE GEODESICS — WHETHER THE SURFACE IS FOLDED OR NOT!

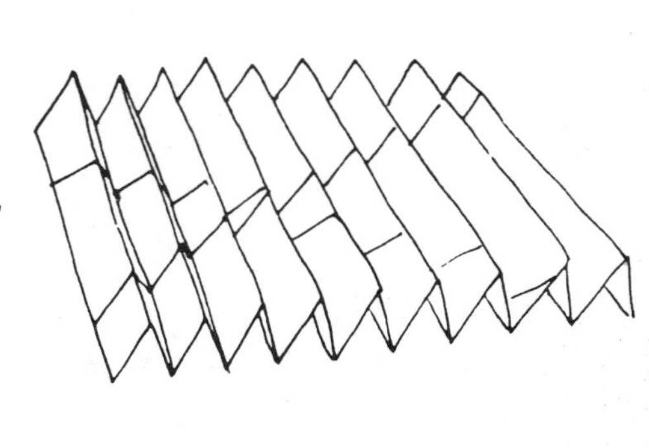

BUT THIS FIRST PART OF OUR JOURNEY IS A FEEBLE THING INDEED, COMPARED TO THE NEXT STEP:

THREE-DIMENSIONAL CURVED SPACES

LEMME **OUT!**

MR. HIGGINS ?
HIM
I'M A SALES REPRESENTATIVE FROM EUCLID & CO. I'M TOLD YOU'VE BEEN HAVING PROBLEMS WITH SOME OF OUR PRODUCTS.
EUCLID & CO.
A
I HAVE WITH ME SOME OF THE LATEST ADDITIONS TO OUR RANGE, WHICH I'M SURE WILL BE ENTIRELY SATISFACTORY.
LET'S SEE THEM, THEN.
THREE DIMENSIONS, THAT'S THE SCENE NOWADAYS. THE GEOMETRY OF TWO DIMENSIONS IS A BIT... OUTMODED...
COSMIC SPACEPROBE FLUID
THE LATEST THING IN GEODESICS...
... IS MADE FROM RIGID RODS, WHICH FIT PERFECTLY TOGETHER.
THESE WILL **NOT** ALLOW YOU TO DEVIATE TO THE LEFT OR THE RIGHT ; OR UP OR DOWN – BUT ONLY TO GO **STRAIGHT AHEAD** !

FOR MEASURING AREAS, WHY NOT TRY OUR NEW PAINT? 100 gm. PER SQUARE METER, EXACTLY.
CUNNING!
AND FOR VOLUMES, THIS CYLINDER OF GAS. YOU CAN READ THE VALUE DIRECTLY ON THE METER ATTACHED TO OUR SPACEPROBE.
AND REMEMBER – AREA OF SPHERE $4\pi l^2$, VOLUME $\frac{4}{3}\pi l^3$.
GOTCHA.
EUCLID & CO.
THIS IS A VERY DEMANDING PROFESSION.
THIS TIME ARCHIE LANDED IN A THREE-DIMENSIONAL SPACE. WE CONTINUE TO FOLLOW HIS EXPLOITS...

BEAUTIFUL PRECISION ENGINEERING. AND THE RODS ARE EXACTLY A METER LONG!
BUT, AFTER STICKING TOGETHER A GOODLY NUMBER OF RODS...
OH LORD, HERE WE GO AGAIN!
MY GEODESIC'S CLOSING UP.
A THREE-DIMENSIONAL CLOSED SPACE?
THAT'S REALLY TORN IT!!
ARCHIE, WHO HAD SAT DOWN ON A PASSING ASTEROID TO EAT A FEW SANDWICHES, DECIDED TO SIZE UP THE ANGLES AGAIN.
I'LL CONSTRUCT A TRIANGLE FROM THREE GEODESICS, LIKE I DID BEFORE...
EUCLID & CO.

?!?
I'VE FITTED MY GEODESICS TOGETHER PROPERLY — BUT THE ANGLE-SUM OF MY TRIANGLE IS MORE THAN 180° !
OH WELL...
SPACEPROBE
FSCHHHHHHHH
I'LL MAKE ONE, AND MEASURE ITS VOLUME AND SURFACE AREA.
A SPHERE OF RADIUS ℓ IS THE SET OF POINTS SITUATED A FIXED DISTANCE ℓ FROM A GIVEN POINT, WHICH I'LL CALL N.
THE AREA IS LESS THAN $4\pi\ell^2$...
... AND THE VOLUME IS LESS THAN $\frac{4}{3}\pi\ell^3$!
I'VE HAD QUITE ENOUGH OF THIS.

ARCHIE FURTHER INCREASED THE RADIUS OF HIS SPHERE...
FSSSCHHHHH...
SPACEPROBE
PLANE
AND NOW MY SPHERE HAS BECOME FLAT!
MORE AND MORE...
NOW ITS CONCAVITY HAS REVERSED DIRECTION!
I DON'T UNDERSTAND ANY OF THIS.
PLANE
EUCLID & CO.
A LITTLE LATER...
HELP! THE WALL'S CLOSING IN ON ME!
FSSCHHHHHHH
EUCLID
QUICK, CUT OFF THE GAS!

OOF!

SO... SIMPLY BY BLOWING UP A BALLOON IN A THREE-DIMENSIONAL SPACE, HIGGINS FOUND HIMSELF — **INSIDE IT!**

IF HE HADN'T TURNED OFF THE GAS IN TIME, HE WOULD HAVE BEEN UTTERLY CRUSHED, IN JUST THE SAME WAY HE ENDED UP TRAPPED IN HIS OWN ENCLOSURE ON PAGE 13.

WITH THE BEST WILL IN THE WORLD, IT'S NOT REALLY POSSIBLE TO **VISUALIZE** THE **CURVATURE** OF THIS THREE-DIMENSIONAL SPACE. ITS GEODESICS CLOSE UP, AND ITS TOTAL VOLUME IS A **FINITE** NUMBER OF CUBIC METERS, LIKE THE SURFACE OF OUR PLANET, WHICH OCCUPIES ONLY A FINITE NUMBER OF SQUARE METERS.

THE ANGLE-SUM OF A TRIANGLE, IN THIS THREE-DIMENSIONAL SPACE, IS MORE THAN 180°. TO "**SEE**" THE CURVATURE YOU WOULD HAVE TO BE ABLE TO ENVISAGE IT IN FOUR DIMENSIONS.

IT COULD BE TRUE THAT OUR THREE-DIMENSIONAL **UNIVERSE** IS A **HYPERSURFACE** IMMERSED IN A FOUR-DIMENSIONAL SPACE, WHICH IS ITSELF IMMERSED AS A HYPERSURFACE IN FIVE-DIMENSIONAL SPACE, AND SO ON. BUT, AT PRESENT, IT ISN'T CONSIDERED GOOD TASTE TO DISCUSS SUCH MATTERS...

WHAT WOULD THE WORLD BE COMING TO, WITH IDEAS LIKE THAT, I ASK YOU?

WHAT EXISTS IS WHAT I CAN SEE!

EVERYFINK ELSE IS JUST... METAPHYSICS!

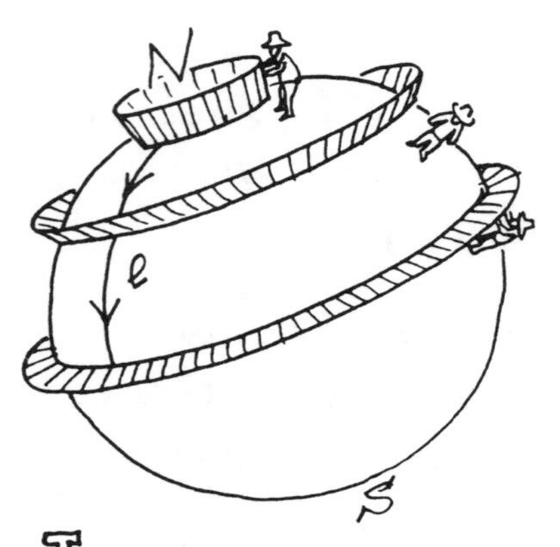

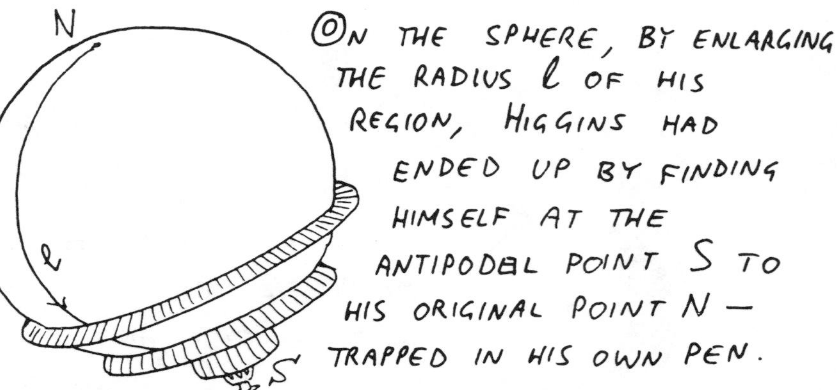

On the sphere, by enlarging the radius ℓ of his region, Higgins had ended up by finding himself at the antipodal point S to his original point N — trapped in his own pen.

In a 3-dimensional space of positive curvature, the same thing happens. In his 2-dimensional sphere, Archie reached the **EQUATOR**, enclosing half the available area. In this 3-dimensional **HYPERSPHERICAL** space, there is an **EQUATOR** too; and Archie reached it when his balloon occupied half the available volume. On the sphere, the equator looked like a **STRAIGHT LINE**. Likewise, on the hypersphere, the "equatorial balloon" looked like a **PLANE**.

After passing the equator the concavity of the balloon reversed, and he moved automatically towards the point S antipodal to N, the center of the balloon.

On a sphere, every point has an **ANTIPODE**. It's just the same on a hypersphere in 3 dimensions – even though it's a little difficult to grasp immediately.

PROBLEMS?
WELL - UH - IT'S ALL GOT MUDDLED UP A BIT INSIDE MY HEAD.
I'M CALLED SOPHIE. CURVES OF ALL KINDS - THAT'S MY LINE.
NAVIGATION ON A HYPERSPHERE IS A LITTLE SURPRISING AT FIRST. THE BEST WAY NOT TO GET STUCK IS TO TAKE IT A LITTLE AT A TIME.

MMM, YES.
I'VE LOST THE THREAD A BIT.

THE CENTER OF A HYPERSPHERE, HAVING 3 DIMENSIONS, CAN BE FOUND IN A 4-DIMENSIONAL SPACE, PROVIDED WE ASSUME IT IS SO **IMMERSED**. BUT IT DOESN'T LIE ON THE ACTUAL HYPERSPHERE. SIMILARLY YOU CAN IMMERSE A 4-DIMENSIONAL HYPERSPHERE IN A 5-DIMENSIONAL SPACE, AND SO ON AS FAR AS YOU LIKE...

N.
OK THEN - REMEMBER YOURSELF FLATTENED OUT LIKE A LITTLE LABEL ON YOUR 2-DIMENSIONAL WORLD...
... AND YOU BEGAN TO INFLATE YOUR CIRCLE - WHICH IS JUST A SPHERE OF DIMENSION ONE...
N
...IN TWO-DIMENSIONAL SPACE, A FRONTIER CONTAINS A SURFACE. SIMILARLY, IN A SPACE OF THREE DIMENSIONS, A FRONTIER IS THE BOUNDARY OF A VOLUME.
A
AHA! THAT'S WHEN I GOT TO THE HALF WAY MARK AT THE EQUATOR.
IN A 4-DIMENSIONAL SPACE, A FRONTIER WOULD HAVE 3 DIMENSIONS, AND BE THE BOUNDARY OF A HYPERVOLUME HAVING 4 DIMENSIONS.
OH 'ECK, 'E'S ORF AGAIN!
LET'S BEAT IT!
N
LOOK - HERE'S YOUR CIRCLE, A "ONE DIMENSIONAL BALLOON". IT'S STARTING TO COVER MORE THAN HALF OF THE AVAILABLE SPACE - CLOSING IN ON ITSELF, AND ON YOU, AND CONVERGING TOWARDS THE ANTIPODE S.
S

JUST THE SAME AS, IN MY CURVED 3-DIMENSIONAL SPACE, ONCE I'D PUMPED IN MORE THAN HALF THE TOTAL VOLUME, THE BALLOON CLOSED IN, ON ME, HEADING TOWARDS THE ANTIPODAL POINT.
I'VE UNDERSTOOD IT!
SINCE THE SPHERE, IN THIS 3-DIMENSIONAL CURVED SPACE, OBVIOUSLY HAS **TWO** CENTERS, WHICH ARE ANTIPODAL.
?!!?
WELL, THAT IS... I'M NOT EXACTLY SURE **WHAT** I'VE UNDERSTOOD, BUT I HAVE THE IMPRESSION I'VE UNDERSTOOD **SOMETHING**.
HOW DEPRESSING!
THAT'S O.K, ARCHIE. IN MORE THAN THREE DIMENSIONS, **TO UNDERSTAND IS TO EXTRAPOLATE.**
I'M EXTRAPOLATING, BUT I DON'T UNDERSTAND!
YOU HAVE TO BUILD THE PICTURE YOURSELF, IN YOUR IMAGINATION!

NOW, I'LL START WITH A 3-DIMENSIONAL SPACE AND PUT A LOT OF SPHERES – TINY 2-DIMENSIONAL UNIVERSES – INSIDE IT.

THESE UNIVERSES CAN INTERPENETRATE. THEIR COMMON POINTS FORM CIRCLES – OBJECTS OF DIMENSION **ONE**.

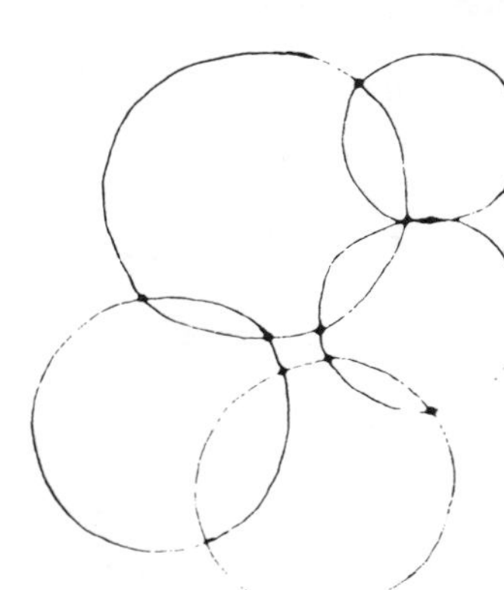

SIMILARLY THESE CIRCLES, HAVING A SINGLE DIMENSION, WHEN PLACED ON A SHEET OF PAPER (DIMENSION 2) CUT IN **POINTS**. (IT IS CUSTOMARY TO SAY THAT THE DIMENSION OF A POINT IS **ZERO**.)

SO A SPHERE CAN BE VIEWED AS THE INTERSECTION OF TWO 3-DIMENSIONAL "BUBBLES" LIVING IN A SPACE OF 4 DIMENSIONS.

AND SO IT CONTINUES: A 3-DIMENSIONAL CURVED SPACE, A HYPERSPHERE, CAN BE THOUGHT OF AS THE INTERSECTION OF TWO 4-DIMENSIONAL SOAP-BUBBLES IN A SPACE OF 5 DIMENSIONS.

ARCHIBALD AND SOPHIE, HAVING SCALED SUCH GIDDY HEIGHTS OF EXTRAPOLATION, RECOMMENCE THE EXPLORATION OF NEW 3-DIMENSIONAL WORLDS.
MATHEMATICS IS NO MORE THAN WHAT IT IS... ISN'T IT?
THIS, OF COURSE, IS 3-DIMENSIONAL STICKY TAPE, FOR MAKING GEODESICS. THE STICKY BIT'S ON THE BOTTOM, NATURALLY.
AND NOW, IN THIS NEW SPACE, GEODESICS DON'T CLOSE UP. AND WHEN I INFLATE THE SPACEPROBE BALLOON, THE VOLUME USED IS GREATER THAN $\frac{4}{3}\pi \ell^3$; AND THE SURFACE AREA IS MORE THAN $4\pi \ell^2$. FURTHER, THE ANGLE-SUM OF A TRIANGLE IS LESS THAN 180°.
RECALLING PAGE 23, YOU'LL REALIZE YOU'RE IN A SPACE OF **NEGATIVE CURVATURE**!

RÉSUMÉ:

IN THREE-DIMENSIONAL SPACES, THERE ARE LOTS OF POSSIBLE KINDS OF BEHAVIOR, YOU KNOW. IT'S JUST LIKE YOU GET WITH SURFACES, WHICH ARE TWO-DIMENSIONAL SPACES.

IF THE ANGLE-SUM OF A **TRIANGLE**, IN A 3-DIMENSIONAL SPACE, IS GREATER THAN 180°, THEN WE SAY THAT THE **CURVATURE** IS **POSITIVE**. THEN, FORMING A SPHERE OF RADIUS ℓ, THE **SPACEPROBE** GIVES A VOLUME LESS THAN $\frac{4}{3}\pi\ell^3$ AND AN AREA LESS THAN $4\pi\ell^2$. THIS SPACE, A **HYPERSPHERE**, CLOSES UP ON ITSELF. BUT, IF THE ANGLE-SUM OF A TRIANGLE IS LESS THAN 180°, THEN THE CURVATURE OF THE 3-DIMENSIONAL SPACE IS **NEGATIVE**. THE VOLUME OF A SPHERE OF RADIUS ℓ IS MORE THAN $\frac{4}{3}\pi\ell^3$ AND ITS SURFACE AREA IS MORE THAN $4\pi\ell^2$. THE WHOLE SPACE IS OF **INFINITE** EXTENT.

BUT IF THE ANGLE-SUM COMES TO 180°, THE SPACE IS SIMPLY **EUCLIDEAN**.

IS **THAT** WHAT WE'VE GONE THROUGH ALL THIS FOR? PAH!

A SPACE MUST BE EITHER OPEN OR CLOSED!

WHAT'S HAPPENED NOW - AND WHERE THE DEVIL AM I ?
SOPHIE ?
?
?
AHOY!
AHOY!
???
???

BACK SOON
BACK SOON
BACK SOON
BACK SOON
BOGGLE

YOU SEE – HIGGINS WAS INSTALLED IN A **CYLINDRICAL** 3-DIMENSIONAL SPACE.

DESPITE BEING EUCLIDEAN, WITH ZERO CURVATURE (ANGLE-SUMS ARE 180°) THIS UNIVERSE CLOSES UP ON ITSELF.

H'OKEY-DOKEY! WE GOT SPHERICAL SPACES, 'YPERBOLICAL ONES, AN' SEE-LINDRICAL ONES. THAT'S THE LOT, AIN'T IT ?

YOU THINK SO ?

LET'S TAKE A LITTLE TRIP BACK TO TWO DIMENSIONS.

INSIDE OUTSIDE:

IT'S CLOSED UP, JUST LIKE IT DID BEFORE.
ANGLE-SUM 180° THIS GADGET IS EUCLIDEAN. ANOTHER CYLINDER!
OR MAYBE NOT...
!?!
THAT'S FUNNY...
START OF THE GEODESIC
ALL IS NOW REVEALED... THIS TIME ARCHIE WAS IN A **NON-ORIENTABLE 2-DIMENSIONAL SPACE**. THE BEST KNOWN EXAMPLE OF SUCH A SPACE IS THE MÖBIUS BAND (1830). THE IDEA HAD ESCAPED THE GREEKS, EVEN THOUGH THEY THOUGHT ABOUT EVERYTHING.
START OF THE GEODESIC
A
SOPHIE? LET ME OUT!

DRAW A CIRCLE ON A SURFACE, AND PUT AN ARROW ON IT. THINK OF THE CIRCLE AS A LITTLE LABEL WHICH WE CAN SLIDE AT WILL OVER THE SURFACE. IF THE CIRCLE ALWAYS RETURNS TO ITS ORIGINAL POSITION WITH THE ARROW POINTING THE SAME WAY, WE SAY THAT THE SURFACE IS **ORIENTABLE** – AS IS THE CASE FOR THE SPHERE, CYLINDER, PLANE, ETC. BUT ON A MÖBIUS BAND, THINGS GO QUITE DIFFERENTLY...

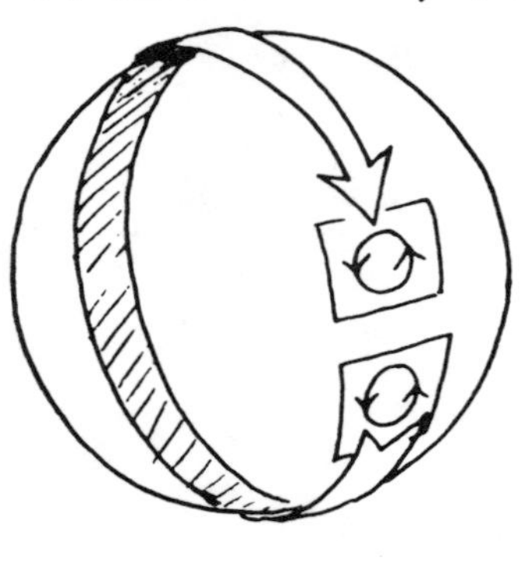

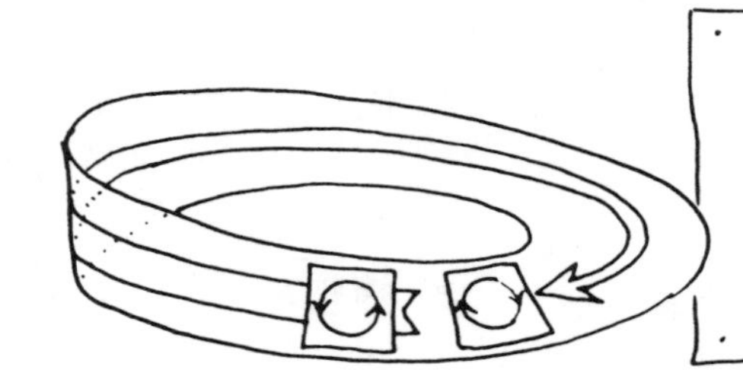

EVERY TIME IT TRAVELS ROUND THIS 2-DIMENSIONAL UNIVERSE, THE CIRCLE REVERSES ITS ORIENTATION.

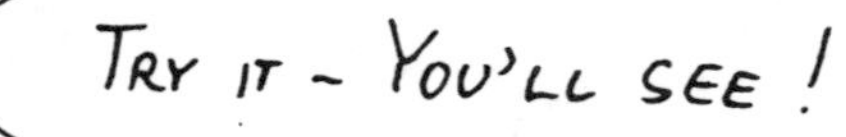

IN THE SAME WAY, YOU CAN'T PAINT THE MÖBIUS BAND WITH A DIFFERENT COLOR ON EACH SIDE: IT HAS ONLY **ONE** SIDE! WE SAY IT IS **UNILATERAL**.

IT HAS ONLY ONE **EDGE**.

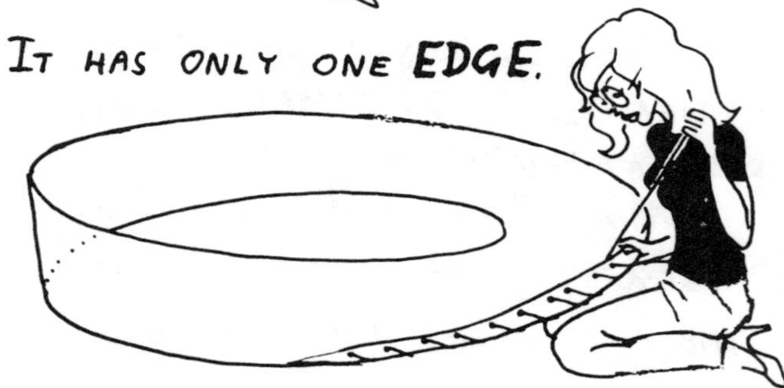

YOU CAN HEM IT ALL AT ONCE

O.K, LET'S TRY CUTTING IT IN TWO
EASIER SAID THAN DONE, ARCHIE MY FRIEND.
THEN HOW DO YOU CUT IT IN TWO?
EASY. YOU CUT IT IN THREE.
NOTE THAT THE GADGET HAS NOW BECOME TWO-SIDED (OR BILATERAL)
NOW I'M GETTING DISORIENTED.
NOTICE THAT WE NOW HAVE A ONE-SIDED THINGY (WHITE) AND A TWO-SIDED THINGY (GREY) WHICH IS TWICE AS LONG AS THE ORIGINAL STRIP.

AFTER THIS BRISK TROT AROUND THE MÖBIUS BAND, LET'S GO BACK AND TAKE ANOTHER LOOK AT 3-DIMENSIONAL EUCLIDEAN SPACES.

THE ORIENTATION OF SPACE:

LET US ACCOMPANY HIGGINS ON HIS EXPLORATION OF ANOTHER EUCLIDEAN WORLD OF THREE DIMENSIONS (WITHOUT CURVATURE)
WHERE'S SOPHIE ? WE'D ARRANGED TO MEET FOR LUNCH ON PAGE 57.
THIS SPACE HAS AN AIR OF STRANGENESS WHICH DOESN'T BODE WELL.
I'LL HAVE A DRINK WHILE I'M WAITING.
JUST LOOK AT THE FOG CREEPING IN. AND ME WITHOUT MY ASPIRIN !
! OOH-OOY
EH ?
WHERE ARE YOU ? COME OUT INTO THE OPEN !

HI! HI! HI!
HI! HI! HI!
I'LL MAKE A FEW TESTS.
ANGLE-SUM 180°...
AREA $4\pi l^2$ VOLUME $\frac{4}{3}\pi l^3$
IT'S EUCLIDEAN.
THAT'S LUCKY. THERE'S MY BOTTLE.
WHAT'S THE MATTER WITH THIS @#**!!!***@!! CORKSCREW ?
IT'S **MY** CORKSCREW ALL RIGHT! I RECOGNIZE IT.
I WONDER WHAT'S HAPPENED TO IT ?
TAKE A CLOSE LOOK!
TROUBLE, SPORT ?
I — UH...

FAIR DINKUM, IT'S A BLINKIN' FRENCH ROSÉ! IF THAT DON'T BEAT ALL!
ROSE DE PROV
Y'GOT A CORKSCREW, COBBER?
UH-YES...
WALTZING MATILDA; WALTZING MATILDA; YOU'LL COME A-WALTZING MATIL-
BLUB BLUB BLUB
YOU DIDN'T NOTICE ANYTHING STRANGE?
I'VE DRUNK BETTER FROM A BILLABONG, BUT THERE'S NOTHIN' STRANGE ABOUT THAT!
TYPICAL O' YOU POMS – ALWAYS ASKIN' BLINKIN' SILLY QUESTIONS...
GIDDAY, COBBER!
GLOP
GLOP
GLOP
GLOP
SOMEONE'S SCREWED UP MY CORKSCREW, AND AN INEBRIATED KANGAROO'S SNITCHED MY ROSÉ. MUCH MORE OF THIS AND I'LL START TO WONDER WHAT'S GOING ON...
DRAT!
SBLING!
WE STRONGLY ADVISE YOU TO OBSERVE THE CORKSCREW WITH THE UTMOST ATTENTION.

THE MÖBIUS BAND – A NON-ORIENTABLE 2-DIMENSIONAL SPACE – HAS A 3-DIMENSIONAL ANALOG.

ON A MÖBIUS BAND, A CIRCULAR LABEL THAT MAKES A "CIRCUIT" IN THE SPACE, CAN COME BACK WITH ITS ORIENTATION CHANGED.

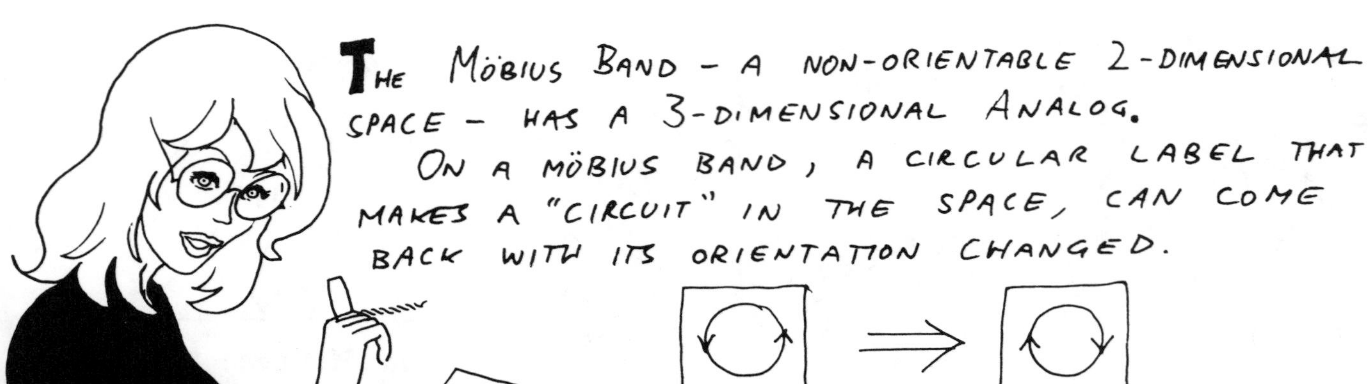

SEE PAGE 54

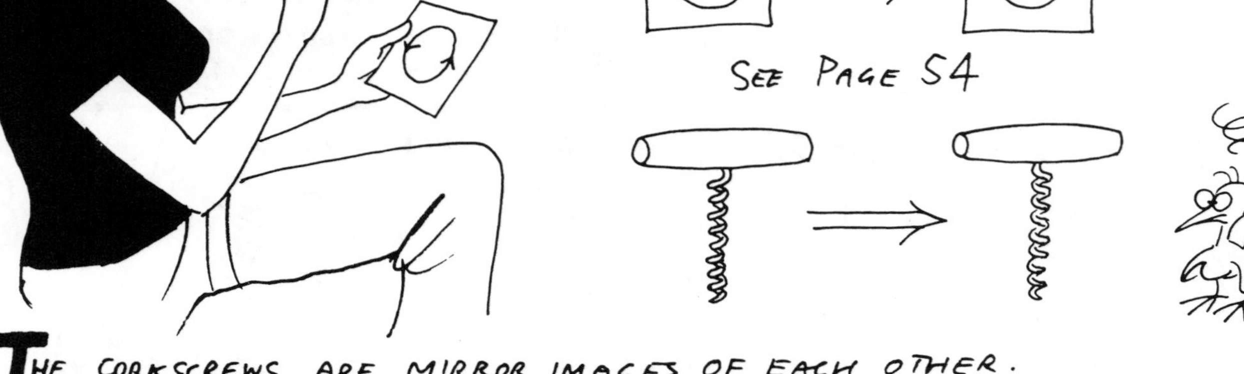

THE CORKSCREWS ARE MIRROR IMAGES OF EACH OTHER.

THE CORKSCREW, AND ARCHIE HIMSELF, CAN BE THOUGHT OF AS "LABELS" IN THREE DIMENSIONS. EACH TIME AN OBJECT MAKES A "CIRCUIT" OF THIS 3-DIMENSIONAL SPACE, ITS ORIENTATION REVERSES. AS WE ACCOMPANIED HIGGINS ON HIS CIRCUMSPATIAL PILGRIMAGE, IT'S NOT SURPRISING THAT, JUST LIKE HIM, WE FOUND THE BOTTLE TO BE A MIRROR IMAGE, AND THE CORKSCREW TWISTING THE WRONG WAY. A SECOND "CIRCUIT" WOULD RESTORE THESE OBJECTS TO THEIR ORIGINAL APPEARANCE, PROVIDED WE LEFT THEM WHERE THEY WERE.

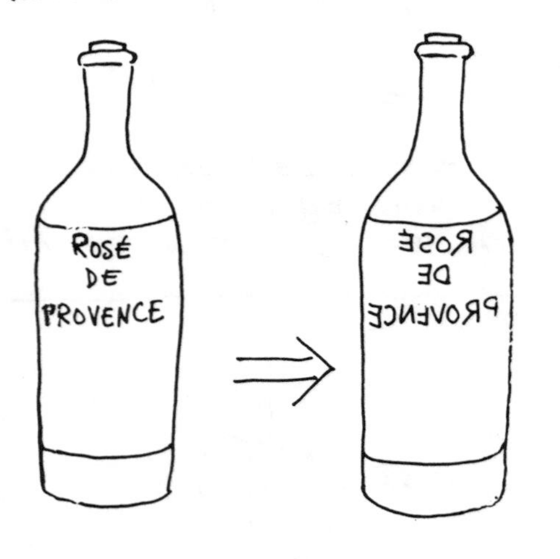

ARCHIE AND THE KANGAROO (AN ANTIPODAL SPECIES) LIVED IN THE SAME SPACE; BUT THEY DIFFERED IN THE SENSE THAT WHAT WAS THE RIGHT WAY ROUND FOR THE KANGAROO, WAS THE WRONG WAY ROUND FOR ARCHIBALD – AND VICE VERSA.

EPILOGUE:
IT'S ALL GONE HAYWIRE. THERE'S NO MORE LEFT OR RIGHT, NO CLOCKWISE OR COUNTERCLOCKWISE, NO RIGHT WAY AND NO WRONG WAY. WHICH WAY, THEN, SHOULD I GO ?
YOU MUST FOLLOW THE GEODESICS, ARCHIE - THE GEODESICS OF YOUR LIFE.
HUH. YER'LL NEVER GET ME TO BELIEVE THE UNIVERSE IS THAT CRAZY! H'IT'S ALL RAVINGS OF A LOONY MATHEMATICIAN!
IT'S LIKE SOMETHING OUT OF A COMIC STRIP!
WHY BOVVER WIV THAT TRIPE, W'EN IT'S BLEEDIN' H'OBVIOUS THAT THE UNIVERSE IS H'EUCLIDDYAN! (*)
(*) A VIEW STATED IN 1830 BY OSTROGRADSKY, A PROFESSOR OF MATHEMATICS AT PETROGRAD, AFTER A LECTURE ON THE WORK OF RIEMANN AND LOBACHEVSKY.

H'IMAGINE THE UNIVERSE DON'T LOOK LIKE WOT IT IS? CRIPES, FINK O' THAT BEIN' TAUGHT IN THE SCHOOLS !!
HOW UTTERLY DREADFUL!
ANY'OW, WOT REELY COUNTS IS THE REAL WORLD – NOT THIS RUDDY H'IVORY TOWER SPECKYERLATION! AN' IN THE REAL WORL' AS YER'LL NO DOUBT H'AGREE WIV ME, YER GONNA
SO, WHAT'S BEHIND ALL OF THIS?
PHYSICS, LOVE...
I'LL BRING THIS INTO THE OPEN!
LET'S HEAR IT FOR THE CONCRETE.
ANYONE THERE?

... IF THE DENSITY OF MATTER IN THE UNIVERSE IS LESS THAN 10^{-29} gm. cm.$^{-3}$...
?

... THEN THE UNIVERSE WILL HAVE NEGATIVE CURVATURE AND INFINITE EXTENT...

... ON THE OTHER HAND, IF THE DENSITY IS GREATER THAN 10^{-29} gm. cm.$^{-3}$
!..

... THE UNIVERSE WILL BE HYPERSPHERICAL AND CLOSED.

AND IF THE DENSITY IS EQUAL TO 10^{-29} gm. cm.$^{-3}$, IT WILL BE EUCLIDEAN...
HMMM

DOCTOR EINSTEIN, I PRESUME?
THE SAME.
THE END

IS THERE A
MATHEMATICIAN
IN THE HOUSE ?